TREVITHI
2001
PROJECT

BUILDING A REPLICA OF THE 1801 CAMBORNE ROAD LOCOMOTIVE

Colin French & Phil Hosken

Camborne 2001

MAP SHOWING THE LOCATION OF KEY PLACES MENTIONED IN THE TEXT AS THEY WERE IN CAMBORNE IN 1801

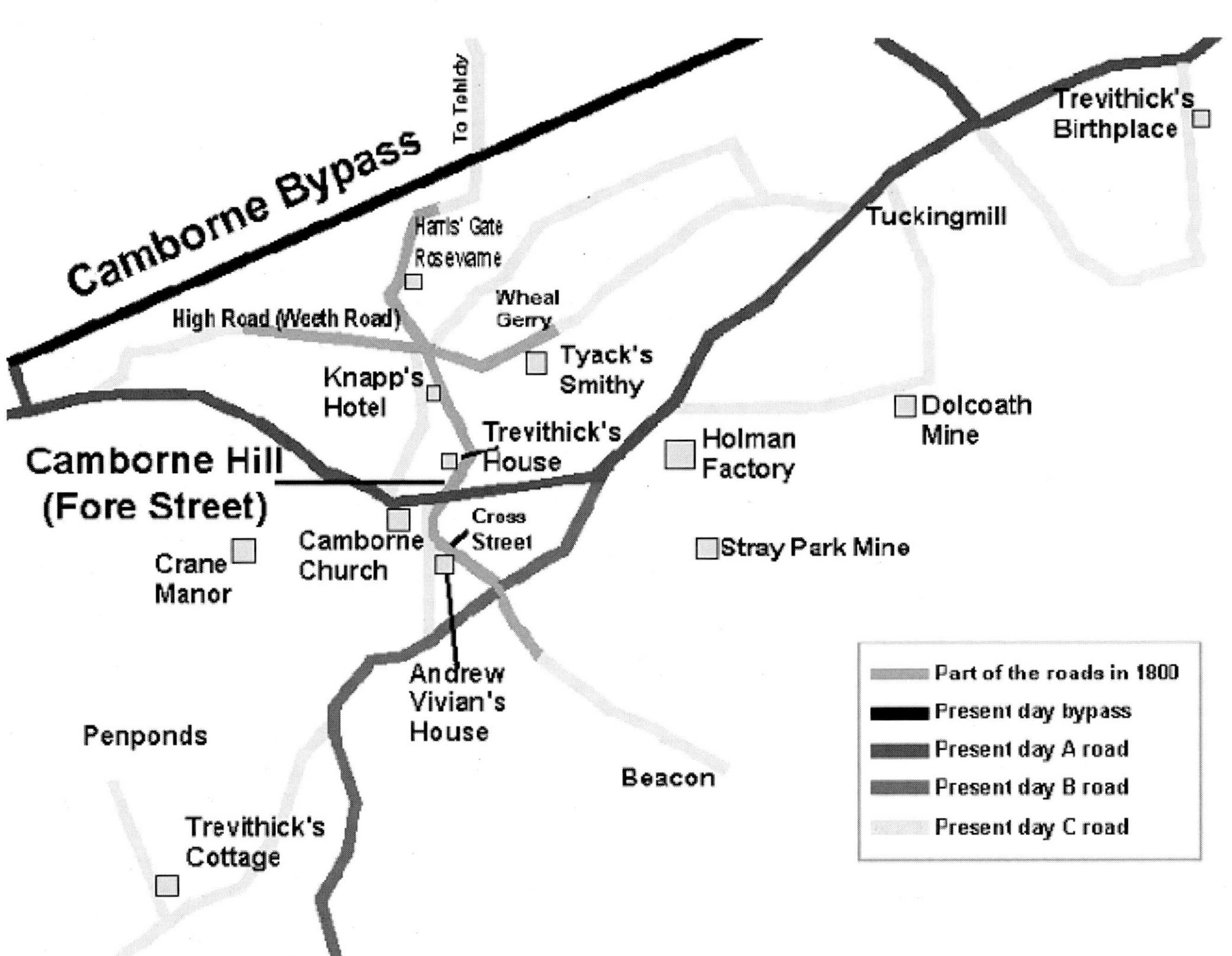

PREFACE

When one considers the many technological achievements of mankind, few people can match those of Richard Trevithick. He was a pioneer of the Industrial Revolution and undoubtedly one of the greatest engineers to have ever lived. The range and magnitude of his inventive genius is truly astounding, and yet few outside Cornwall are aware of the immense contribution he made to the development of the modern world.

In Camborne, on Christmas Eve 1801, one of the most momentous events in the Industrial Revolution took place; an event that was destined to transform the entire economy of the world and affect the lives of countless millions of people. On that date, Richard Trevithick successfully completed the building of the first ever self-propelled, passenger-carrying motorised vehicle and drove it for the first time. A steam locomotive which can be thought of as the first powered automobile or motor car. Indeed, the ancestry of every car and train in the world today can be traced back to that one journey through the streets of Camborne. A defining moment in history, which ranks alongside that of the famous flight by the Wright Brothers or the invention of the telephone by Alexander Graham Bell.

Since 1996, the Trevithick 2001 Project has been the flagship project of the Trevithick Society, a registered charity. The project was to build a full-size steam-powered replica of the Camborne road locomotive ready for its bicentenary in the year 2001. Although the actual date of the bicentenary is Christmas Eve 2001, it was decided at the outset that the replica should be ready for its first public run during the Trevithick Day celebrations in Camborne at the end of April 2001. It would then be involved in a host of events during the year leading up to the actual bicentennial re-enactment on Christmas Eve.

From the beginning this project was seen as much more than simply building a replica. There was a wider view of playing a full and constructive role in the economic regeneration of Cornwall. The Camborne road locomotive is highly symbolic of a time when Cornwall was at the forefront of technology and whilst it is valuable to remember that illustrious past, it is more important to look forward, just as Trevithick constantly did, and have a vision for the future.

If history teaches us nothing else, it demonstrates that given the opportunity and support, the people of Cornwall are fully capable of embracing and developing new technology and being part of a vibrant economy. That support was dramatically taken away in the 1860s when the copper mining industry collapsed and there followed a long, protracted economic decline until the present parlous state of the Cornish economy. Regeneration and substantial investment in the manufacturing, and wealth creating, sector is desperately needed and long overdue. However, to succeed, this regeneration needs to be driven from within, by local people and not managed by outside agencies. Objective 1 should be that vehicle to success and has started to pour money into Cornwall, but, in these early stages, is proving too bureaucratic, is hindered by outside control and is completely missing the target. The basis of a sustainable economy needs to be built on investment in the existing industry that it may develop, expand, become more competitive and embrace new technology. That will provide a platform for new industry and the jobs that will break the cycle of poverty and deprivation in the communities that Objective 1 is supposed to be targeting.

It is time for Trevithick's contribution to the development of the modern world to receive due recognition and for Cornwall to be a beneficiary. In building this replica, it is hoped that a greater understanding of that contribution might be engendered and that it will provide a catalyst for worthwhile economic regeneration and future prosperity.

Colin French,
March 2001

RICHARD TREVITHICK

Richard Trevithick was born on April 13 1771 in the parish of Illogan in a house which still stands alongside Station Road, Pool. Soon after, the family moved to Penponds where he lived until his marriage to Jane Harvey in 1797. This thatched cottage also still exists and is owned by the National Trust. His father, Richard Trevithick Snr., was a man of some repute being the mine manager at Dolcoath, a number of other mines in the neighbourhood, and the mineral agent for Lord De Dunstanville from 1777. Furthermore, he was a devout Methodist and even entertained John Wesley, at Penponds, during his visits to Cornwall.

THE COTTAGE AT PENPONDS

During the last quarter of the eighteenth century, these being the formative years of Richard Trevithick, Cornwall was home to a greater number and concentration of steam engines than anywhere else in the world. These were all Newcomen engines but as his childhood progressed they were increasingly being replaced by the more efficient Watt engines.

Young Richard was not a model schoolchild and his rudimentary education in Camborne, where he was a difficult, inattentive and frequently truant pupil, contributed little to his development. Instead, his education came from living in the midst of one of the most heavily industrialised places in Britain. It was an area at the forefront of technology with Camborne, Redruth and Hayle, in particular, a melting pot of inventiveness meeting the growing demands of the mining industry. At fifteen he was at work at Dolcoath Mine and at nineteen he was appointed engineer at the nearby

A NEWCOMEN ENGINE HOUSE IN CORNWALL

Stray Park mine. No doubt his size helped his meteoric rise, for at 6’ 2” and well built, he was a giant of his age and had legendary strength and a proficiency at Cornish wrestling. However, it was his strength of mind, his practical problem-solving skills and his innovative ability that came to the fore. Indeed, by the age of twenty-one, in 1792, he was sufficiently respected that he was asked to test and report on the performance of a new engine erected at Tincroft. He evidently had an excellent ‘apprenticeship’ in mining and contemporary engineering.

The constant need, in the eighteenth century, to keep the ever deepening mines of Cornwall dry, made it necessary to employ the beam engines of Newcomen and latterly the more efficient engines of Boulton and Watt. They were monstrous stationary machines which were very expensive to

A BOULTON AND WATT ENGINE

erect, each requiring an immense masonry engine house and stack, a separate boiler and a pond to maintain a constant water supply.

Boulton and Watt held very restrictive patents which effectively gave them a monopoly on the supply of steam engines for much of this period. In Cornwall there were many unsuccessful experiments and trials made to circumvent these patents and it is likely that this quest was a driving force which propelled the inventive mind of the young Richard Trevithick. Certainly he became embroiled in many of these experiments and incurred the litigious and venomous wrath of James Watt.

It was during this time that Richard first met Davies Gilbert (Giddy) who was to become his lifelong mentor and friend. Davies Gilbert, another Cornishman, was an eminent mathematician and academic and went on to be the President of the

HAZELDINE
TREVITHICK PUFFER
SCIENCE MUSEUM

CUGNOT'S STEAM TRACTOR

Royal Society. When considering the prospect of a non-condensing high-pressure engine to avoid Watt's patent, Gilbert's advice was sought. In 1839, Gilbert wrote "On one occasion Trevithick came to me and enquired with great eagerness as to what I apprehended would be the loss of power in working an engine by the force of steam raised to the pressure of several atmospheres, but, instead of condensing, to let the steam escape. I, of course, replied at once that the loss of power would be one atmosphere", and then commented "I never saw a man more delighted".

In turn, Trevithick developed this notion and revolutionised the concept of the steam engine in an act of miniaturisation probably not equalled until the advent of the silicon chip. In these final few years of the eighteenth century, he successfully pioneered the use of high pressure steam and made a huge advance in technology by developing a new type of powerful compact steam engine where the component parts of the engine were placed within the boiler. Thus, he invented a cheap, eminently portable, non-condensing engine as powerful as the contemporary leviathans of Newcomen or Boulton and Watt. A reduction in size from about 1000 tons to under 20 tons, when one considers the masonry enginehouse, pond, etc. which were integral parts of the Watt engine.

This revolutionary design, and developments of it, formed the power source that transformed the industrialised world during the nineteenth century. Such was the versatility and relative cheapness of this engine that it was used as a stationary engine in factories and farms to operate machinery, as well as the engine on ships, and that of road and rail locomotives. Indeed, the ancestry of every car, train and self-propelled ship in the world today can be traced back to that remarkable invention.

Before Trevithick there were several independent attempts to harness steam power to create self-propelled vehicles. The first of these

REPUTEDLY TREVITHICK'S "FIRST" MODEL IN THE GUINNESS COLLECTION, STRAFFAN NEAR DUBLIN

TREVITHICK'S "SECOND" MODEL IN THE SCIENCE MUSEUM, KENSINGTON

was 1769 in France when N. J. Cugnot built a prototype gun tractor designed to tow artillery. The following year an improved version made one very short journey and as it could not be effectively steered, it careered into a wall and fell over, and Cugnot was promptly arrested. Much closer to home, William Murdoch made an experimental model locomotive in 1784, which was tried out on the streets of Redruth. His employers, Boulton and Watt, deplored what they saw as a waste of their employee's time and if further models were built the work was conducted in secrecy at Budge's brass foundry at Tuckingmill. Such models would have reached, at best, motorcycle-size. None of these remarkable achievements can be described as real progenitors in the development of the steam locomotive and should be considered as evolutionary lines that became extinct. Richard Trevithick was thirteen years old in 1784 and it is not known whether he was aware of Murdoch's trials, although Murdoch's son contended that it was shown to Richard. Nevertheless, even if he did know of those experiments, they certainly did not influence his radically different design: a true testament of Trevithick's own creative genius.

In 1797, the year of his marriage to Jane Harvey and the death of his father, Richard began to develop his ideas for the use of high pressure

LONDON ROAD CARRIAGE

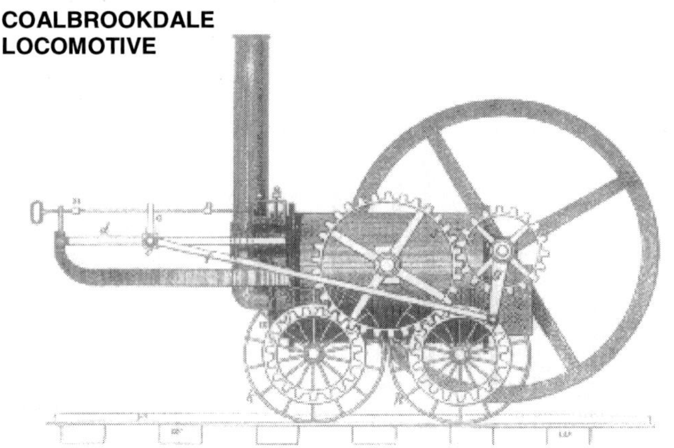
COALBROOKDALE LOCOMOTIVE

steam. He started by using models to test the practicality of using this inherently dangerous and invisible power source. These tests included a model stationary engine and two model locomotives, which are still in existence. These are believed to have been made by William West, Richard's brother-in-law, and demonstrate that, from the outset, Richard realised the potential for motorised transport. The following year he successfully manufactured and operated stationary high pressure engines and realising that his stationary 'puffer' engine was indeed powerful enough to propel itself, soon began to develop the world's first self-propelled passenger carrying vehicle - the first car! At the age of 29, in November 1800, he was ready to build, test and prove his vision of motorised transport, which it must be remembered he financed himself. This was achieved with a group of talented and skilled friends and relations in Camborne, using the most rudimentary of manufacturing equipment. In a remarkably short time, it led to that epic journey 'Up Camborne Hill' on Christmas Eve 1801 (see next chapter for further details) .

With his partner Andrew Vivian, Richard went on to patent his ideas for high pressure steam in March 1802. This patent included an illustration of his next road locomotive, the London Road Carriage. This engine was probably manufactured at the Harvey's foundry at Hayle, where the boiler was made and the iron chassis forged. It is also likely that it was driven around in Cornwall before being shipped to London where the body was added at the Felton coachworks at Leather Lane in Clerkenwell. When fitted out, the first London Bus undertook a number of early morning runs of several miles each through the streets of London. It was estimated that it attained speeds of 8 or 9 miles per hour in the empty streets. It is not certain whether it was a three or four wheel vehicle but it is known to have provided an uncomfortable ride sitting so high above the road.

A second locomotive was begun in 1802 and on 22nd August 1802, Richard wrote a letter to Davies Gilbert ending "… the Dale Company have begun a carriage at their own costs for the realroads [sic] and are forcing it with all expedition". It was designed to run on the 3' gauge plateway tracks and it was also in operation during 1803. It is likely this engine was conceived as a portable power plant to be moved around the works on the plateway system as needed.

THE CATCH-ME-WHO-CAN PASSENGER TRAIN

ONE SHILLING TICKET

The concept of the railway locomotive had to wait until 2nd February 1804. Richard was installing a large puffer to work the rolling mills at the Penydarren Iron Works, in Wales. In extolling the efficiency of Trevithick's high pressure engines, the proprietor, Samuel Homfray, made a bet with Richard Crawshay of the Cyfarthfa Iron Works of 500 guineas, each way, that one of his engines, mounted on wheels, would haul ten tons of iron down the tramway from Penydarren to Abercynon, a distance of over 9 miles, and return unladen. On the above date the engine completed the journey with the additional weight of 70 people and reached speeds of a fast walking pace.

In 1805 a copy of the Penydarren engine was made for the Wylam Colliery, Gateshead. This had flanged wheels to suit the wooden tramway there. Unfortunately, the rails were not strong enough to support the engine and this experiment was short-lived.

One last attempt to promote mechanised transport was made in 1808 and again he turned to London to stage it. The Catch-me-who-can engine was made at Hazeldine foundry at Bridgnorth, Shropshire. This was to become the world's first fare-paying passenger railway and ran on an enclosed circular track near Gower Street for about three months, but again suffered from broken and sinking rails.

Unfortunately, Richard was decades ahead of his time and the technology of making rails was not sufficiently advanced to withstand the weight of his locomotives. Nor was the world ready for locomotion. All was not lost, sales of the puffers were booming and his locomotives were converted into stationary engines. Henceforward his attention turned to many other mechanical and civil engineering feats, including, in 1808, a tunnel under the Thames which almost succeeded, and steam-powered Thames dredgers.

He continued to develop steam technology and made great strides in boiler design, creating the Cornish boiler and then the multi-tubular boiler which Robert Stephenson developed into the locomotive boiler as used in the 'Rocket'. He is also accredited with inventing the Cornish engine.

New horizons beckoned as a result of meeting Francisco Uvillé in 1813. He had come from Peru to source engines for silver mines high in the Andes. He placed a large order with Trevithick, which enticed the great man to Peru in 1816 after there had been installation difficulties. During his eleven years in South America there were sufficient larger-than-life adventures, and fortunes won and lost, to fill a book on their own.

He returned penniless, but with the attitude that he would carry on where he left off, as if he had only been on a weekend tea treat. And so, at Harveys, at Hayle, he began developing new ideas, including small portable multi-purpose engines and the closed-cycle engine for shipping which would reduce fresh-water consumption and facilitate transatlantic journeys. In 1830 he fell out with the Harveys and stormed off to London. His last few years were spent at J. & E. Hall Ltd., Dartford where he designed 150 p.s.i. boilers and experimented with jet propulsion. He died on 22nd April 1833 at the Bull Inn in Dartford and is buried in Dartford Churchyard.

Richard Trevithick was a genius in the field of mechanical and steam engineering whose inventions proved to be of immense importance in shaping the development of the industrialised world during the 19th century. In fact he was much more than a steam engineer. He was a tremendous proponent of the use of iron and he devised an enormous number of new and exciting ways in which iron could be utilised. It was as if iron was a newly discovered substance that was being tried and tested to see what practical use could be made of it. These ranged from the kibbal (used in mines to haul ore to surface), water tanks, cargo containers on ships, buoys, iron ships and masts, floating docks, diving bells, portable room heaters and even a 1000' cast-iron tower to commemorate the passing of the Reform Bill in 1832.

THE TREWITHEN THRESHING ENGINE, SCIENCE MUSEUM

During his lifetime he made and lost many fortunes and was, at times, by today's standards, a multi-millionaire and at one stage bankrupt. His never-say-die attitude, innate optimism, intellectual drive, indomitable spirit and fiercely independent mind carried him through the many seemingly intractable difficulties he faced and drove him ever onwards to innovate and achieve. A truly remarkable man indeed!

GOING UP CAMBORNE HILL

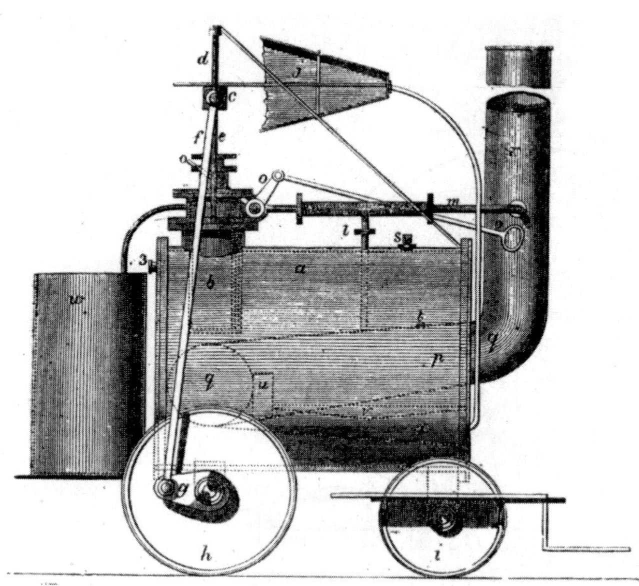

Here's a toast to Cap'n Dick,
And his undoubted skill,
He built a car, the first by far,
And drove up Camborne Hill;
J.S.

The Camborne road locomotive had a very short but successful life, during which time it is known to have made three journeys. These proved the potential for transportation and spurred Trevithick on to further develop his ideas for locomotion in the years to come. The first run on Christmas Eve has the hallmark of impetuosity. The weather was dreadful and dusk was fast approaching, but the engine was ready, and undaunted by the adverse conditions, destiny awaited.

After last minute tinkering the engine was manoeuvred out of the smithy onto the High Road. The boiler fire was lit and slowly the water temperature rose until, approaching an hour, steam began pressurising. Once the desired 60 p.s.i. had been reached, Trevithick, Andrew Vivian and seven or eight others clambered onto the engine, some, no doubt, clinging precariously to the sides. Off she went "like a bird" down Eastern Lane towards the bottom of what is now Tehidy Road and then headed towards the main town, up Fore Street and on to Cross Street, in the general direction of Beacon Hill.

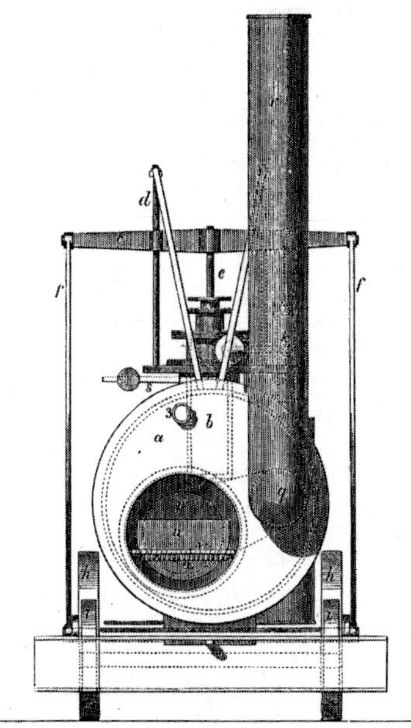

It has often been asserted that the Camborne Hill referred to in the song was Beacon Hill. This is not now thought to be the case and instead Camborne Hill is considered to be the steep stretch of road that is Fore Street. Nevertheless, there are several reports that suggest that the engine did go part of the way up Beacon Hill. Doubt still persists, as that would have been quite an achievement given the gradient and the poor state of any track (it is considered there was no road at that time), and as one wag put it "they would have never made it over the railway crossing".

The following day they took the engine to Crane Manor to visit the relatives of Andrew Vivian. This was the longest and most successful run and yet the least is known about it. Quite

Stephen Williams gave the following report to Francis Trevithick about the first run:

I knew Captain Dick Trevithick very well; he and I were born in the same year. I was a cooper by trade, and when Captain Dick was making his first steam-carriage I used to go every day into John Tyack's blacksmiths' shop at the Weith, close by here, where they put her together.

The castings were made down at Hayle, in Mr. Harvey's foundry. There was a great deal of trouble in getting all the things to fit together. Most of the smiths' work was made in Tyack's shop.

In the year 1801, upon Christmas Eve, coming on evening, Captain Dick got up steam, out in the high-road, just outside the shop at the Weith. When we yet see'd that Captain Dick was a going to turn on steam, we jumped up as many as could; may be seven or eight of us. 'Twas a stiffish hill going from the Weith up to Camborne Beacon, but she went off like a little bird.

When she had gone about a quarter of a mile, there was a roughish piece of road covered with loose stones; she didn't go quite so fast, and as it was a flood of rain, and we were very squeezed together, I jumped off. She was going faster than I could walk, and went on up the hill about a quarter of a mile farther, when they turned her and came back again to the shop.

possibly the engine was christened "puffing devil" on this jaunt, when one elderly lady exclaimed "Good gracious, Mr Vivian, what will be done next? I can't compare un to anything but a walking, puffing devil".

The third and final journey was to be the most audacious and prestigious. It had been arranged that they would visit Lord and Lady De Dunstanville at Tehidy. It is generally accepted that they drove off towards Tehidy but got no farther than Harris' Gate which is an entrance to Rosewarne, where the wheels got caught in a gully, the steering tiller was jerked out of Andrew Vivian's hand and she overturned. The party then manhandled her into a roadside shed and took respite in Knapps Hotel. Davies Gilbert, long after the event, wrote:

"The travelling engine took its departure from Camborne Church for Tehidy on the 28th Dec. 1801 where I was awaiting to receive it. The carriage, however, broke down after travelling well and up an ascent, in all about three or four hundred yards. The carriage was forced under some shelter and the parties adjourned to the hotel and comforted their hearts with a roast goose and proper drink, when forgetful of the engine its water boiled away, the iron became red hot, and nothing that was combustible remained, either of the engine or the house".

Davies Gilbert did not witness the events he described, but his account is interesting for a number of reasons. The start at Camborne Church meant that the engine had to first get to the church and such premeditation indicates this was meant to be a triumphal journey, witnessed by the townsfolk. The description of having travelled "up an ascent, in all about three or four hundred yards" suggests that the engine got much further than Rosewarne, or took another route. As for the dramatic demise of the engine, a more likely interpretation would be that the internal furnace had a fusible lead plug fitted (another of Richard's inventions) which was designed to prevent the possibility of a catastrophic boiler explosion. However, the melting of this safety plug by the overheating engine would have caused a rush of pressurised steam into the flue and burning coal and embers on the fire would have been blown out into the shed catching it on fire.

In 1801 Richard Trevithick lived in Fore Street, Camborne, in a house that conveniently had its own workshop. Richard's cousin, Andrew Vivian, lived in Cross Street, where he also had his own workshop and lathe.

The engine itself was assembled in John Tyack's blacksmith's shop "at the Weith" (see map on page 4). It is now fairly well established

To the Editor of the West Briton.

Sir, - In reading the remarks of "Argus" in last week's West Briton, I am reminded of what my late father, Llewellyn Newton, told me. At the time the engine was made he was with Capt. A. Vivian and Trevithick in their office. Being a good penman and also a good draughtsman, he drew the plans of the engine, under the instruction of Trevithick. The engine was chiefly made in a house inside the walls of Rosewarne. Some of the work was done in a blacksmith's shop, just behind the residence of Capt. A. Vivian, in Cross-street. Our workshop at the present time stands on the same spot where the smith's shop stood, and coal dust can be seen there now. On the first trial of the locomotive it came up through Fore-street on a Christmas-eve at night. My father rode on it with Vivian and Trevithick. The noise it made with the puffing, fire, and smoke caused great fear, and many thought the world had come to an end.

Yours, etc.,
E.T. Newton,
Mathematical Instrument Maker,
Camborne.
12 April 1900

that this blacksmith's shop was at Wheal Gerry beside Eastern Lane, indicating that 200 years ago the Weeth extended further eastwards than we would consider to be the case today (around Weeth School and along Weeth Road). It was constructed by a close knit group of smiths, mechanics, carpenters and coopers from the Camborne area, under the direction of Trevithick himself.

Work began in the autumn of 1800 not long after Richard and Davies Gilbert ran an experiment to prove a horseless carriage could propel itself by friction between the wheel rims and the road. They hired the only available one-horse chaise and manually propelled it up one of the steepest hills around by just applying force to the wheels. This proven production could begin. Some parts, such as the steam cylinder and stuffing boxes, were cast in his father-in-law's foundry, Harvey's Foundry,

GOING UP CAMBORNE HILL,
COMING DOWN,
GOING UP CAMBORNE HILL,
COMING DOWN,
THE HORSES STOOD STILL,
THE WHEELS WENT AROUND,
GOING UP CAMBORNE HILL,
COMING DOWN.

..... WHITE STOCKINGS SHE
WORE,
THE SAME AS BEFORE,
GOING UP CAMBORNE HILL,
COMING DOWN.

Those immortal words commemorate in song the journey of Richard Trevithick's first road locomotive through the streets of Camborne, on Christmas Eve 1801. The words contain a surprising amount of information about the engine and where it went, which leads to the suggestion that the songwriter saw the actual engine. The white stockings would have been the puffs of water vapour from the exhausting steam and "Going up Camborne Hill, coming down" was an observation that it went up hill in reverse with the chimney at the rear.

in Hayle. Other parts were made on a lathe by Captain Andrew Vivian. A steam gauge, wrought iron plate and a barometer came from Coalbrookdale in Shropshire. The set of leather bellows shown in the drawings of the Camborne road locomotive by Francis Trevithick (Richard's son) is known to have been fitted on the original engine, but was of a concertina-type on the cross-head and not as drawn. This set was found to be

ineffective and so was soon removed.

Nicholas Holman established his boiler works in 1801, and tradition has it, made the boiler for the original engine. More likely the boiler shell was cast by Harveys, or possibly Coalbrookdale, and the internal wrought iron boiler fittings were made by Holmans.

Relatively little is actually known about the design of the 1801 Camborne road locomotive. Contemporary accounts are fragmentary and in some aspects contradictory. Francis Trevithick in his biography of his father published some sketches of it but these had no dimensions and were prepared some fifty years after the event, based on the distant memories of others. Francis was an eminent railway locomotive engineer, and at a time when 'Railway was King', his interpretation of what the Camborne road locomotive was like may have been clouded by his desire to see his father remembered as the inventor of the railway locomotive. Indeed, it is possible that the Camborne engine was nothing like the illustrations of Francis. Could such an engine have performed as well as the original did? How could seven or eight people get onboard? Why was the engine patented in March 1802 so radically different? In Francis' favour, the drawings are a logical development of the stationary puffer engine, they are consistent with the models made in the preceding years, and having talked to many people who actually saw the original engine, surely he would have got the fundamental form of the engine correct. We can only speculate but one thing is certain, Francis' illustrations have determined everyone's expectation of the main features and overall layout of the engine, and this has been a guiding principle in the building of the replica.

BUILDING THE REPLICA

Here's a toast to those who came,
And built that car the same,
They built un well, she went like hell,
And climbed the hill again!
J. S.

The idea that the Trevithick Society should find a suitable way to commemorate the bicentenary of the Camborne road locomotive was first mooted in 1992 when John Sawle, a local farmer and traction engine owner, contacted the Society. However, it was not until 1996 that the idea was carried forward, prompted by an e-mail from Karl Petersen, which was published in the February Newsletter. In March 1996 the *Cornish World* magazine carried the suggestion that a commemoration of Trevithick's achievement of 1801 should be celebrated in 2001. Chris Blount, a presenter on BBC Radio Cornwall, took up this suggestion and broadcasted an interview with Philip Hosken, the magazine's editor. Following this John Sawle telephoned to offer his services to build a replica of Trevithick's first road carriage.

After much discussion at Council an investigative committee was formed to examine the possibility of building such a replica. The team was composed of Colin French, Phil Hosken, Charles Thurlow, Courtney Rowe, Ken Brown and Clive Carter. In the beginning John Sawle was not involved but he became so, as soon as it was felt that it would be possible to pursue the proposal as a project.

CLIVE CARTER'S MODEL

Clive Carter, who was to become the energetic chairman of the Society, built a model of Trevithick's 1801 locomotive as depicted by his son Francis in his book about his father. This model proved a useful starting point as it revealed a number of design faults in Francis' drawings that would have made such an engine quite inoperable. The 2001 locomotive was subsequently to benefit from the incorporation of a wooden chassis, fixed man-stand, flywheels, steering and brakes. The set of bellows, as shown incorrectly by Francis, was not to be included as Trevithick himself had noted that it was superfluous, presumably on the first locomotive to be fitted with a blast pipe.

The Society received regular reports on the progress of the committee and it soon became clear that the construction of a replica was a distinct possibility. The Council of the Society encouraged the committee to continue towards a firm objective.

Of course, finance was the great hurdle because such projects require abundant quantities of money, far more than in the reserves of the Society.

ERIC BEST'S MODEL WITH CAP'N DICK PROUDLY AT THE HELM

In the next stage the members of the committee formed themselves into a Project Team and divided their efforts into planning the construction, administration and fund raising.

By this time John Sawle had joined the team and meetings were held on a regular basis. These often lasted until late at night; those attending being sustained by tea and biscuits. The lack of contemporary drawings was a disappointment but not a setback. The protracted search for an illusive Plate XXVI from *A Treatise on The Steam Engine* by Robert Farey proved futile. The original had been deposited at the National Reference Library of Science and Invention in 1851, but was now unavailable. In the event the team relied upon Trevithick's drawings of his other engines, his extant models of 1797 and 1798, and John's visit to the Science Museum at Kensington.

In this latter venture John could have been seen in his stocking feet clambering all over the 1804 Hazeldine industrial *Puffer* engine and Hedley's 1813 *Puffing Billy*, taking photographs and measurements. Notes were also made of contemporary design and construction methods. The Hazeldine engine had been built in accordance with Trevithick's design for a compact industrial steam engine.

Various engineers who had worked on other Trevithick replicas also gave freely of their advice, especially Tom Brogden.

Discussions about the size, working pressure, wheel sizes, speed, fuel consumption and appearance of the finished vehicle brought a remarkable consensus of opinion. The agreement between all those concerned in the initial stages added greatly to the confidence of the Society and everyone who was to have any part in the plan. It soon became clear just how the original carriage would have looked and what had to be done to copy it in the Twenty-first Century. Being the forerunner of the motor car its overall length of 12'0" and width of 5'9" was not going to look incongruous. However its height of 10'6" to the top of the chimney would establish its rightful place amongst the steam engines on Trevithick Day. The other dimensions are the wheelbase of 4'7" and the 4'4" track.

At the end of each meeting Courtney Rowe would return home to make arrangement drawings of the proposed vehicle at that stage. He also had the remarkable ability to calculate the total finished weight of the carriage as the discussions continued about dimensions. The finished running weight is anticipated to be 4.26 tonnes.

The background, aims and methodology of the project had to be encapsulated within a document for presentation to possible funding bodies. Many of these were prepared and dispatched. The response from outside Cornwall was nil. Several large companies had charities to provide funding for such projects as this, but they incorporated various restrictions. One was often the proximity of the project to one of their factories. Factories are not something by which Cornwall is well known today. Approaches to large service industries with branches in Cornwall such as South West Water, together with banks and supermarkets also drew a blank. However, the generosity of the Cornish companies who were approached was excellent and they will always be owed a great debt of gratitude.

On approaching Business Link for support, (who shortly became Prosper), the team were delighted that they were willing to fund the initial costs. These included the building of a 1/12th scale model of the replica by Eric Best. Eric's skill and knowledge of steam vehicles, (he owns a road roller) was a great advantage to the project as he was able to include many smaller parts to his own design. A human touch was supplied by Carn Brea Women's Institute who dressed a model of Action Man as a scale representation of Richard Trevithick.

The model was almost complete when it was shown to HRH The Prince of Wales, Duke of Cornwall, when he visited the Pall Industrial Hydraulics factory at Cardrew, Redruth in 1998. The Duke showed a great interest in the project and a remarkable knowledge of the history of Richard Trevithick.

JOHN TRAVIS, JOHN SAWLE AND DAVID BRAY AT WORK

The original plan, if possible, was to build two replicas: one to remain on show in Cornwall and the other to act as an ambassador of Cornish vision elsewhere in the world. Thus considerable funding was sought for this ambitious project and the intended budget included a road trailer. The building of the replica needed to start at the end of the European Objective 5b funding when all the money intended for such projects had been spent and before Objective 1 funding became available.

An approach to the Government Office for

STEVE HICKMAN, DEEPDALE ENGINEERING, DUDLEY

the South West (GOSW) revealed that there was just £40,000 left in a Regional Development Fund known as Konver II. The team were asked if the replica could be built for that amount plus matching funding. Knowing that all other attempts to raise funds had drawn blanks it was agreed that the task could be completed for that amount, whilst depending very greatly upon other donations and the generosity of manufacturing friends in Cornwall. This funding would be the only chance to get the replica built on time. So, with a very much-diminished budget revised plans were made.

Phase Two of the project began when an Offer Letter from the GOSW was received on 12 November 1999 and work commenced in earnest. The arrival of the Offer Letter also acted as a lever to call down funding from other sources, which had only been promised at that time. These included Kerrier District Council, The West of England Steam Engine Society, Camborne Town Council, The Trevithick Society itself and Mr Frank Trevithick Okuno. A very welcome and substantial donation arrived unexpectedly from the Tanner Trust and it allowed the team to print a progress leaflet for Trevithick Day 2000 and cover some of the unbudgeted expenses that inevitably arose.

John started work on the engineering drawings and negotiations were opened with boilermakers and other important contributors of parts. Deepdale Engineering of Dudley was chosen to build the boiler. A traditional engineering company, they specialised in heavy press work and saw the Trevithick boiler as a lightweight job. Nevertheless they took considerable interest in the project and completed their work by March the following year. Inspections and certification of the boiler to EC pressure vessel standards were carried out at all stages of construction. The working pressure will be from 60-100 p.s.i. Deepdale also made the elbow bend at the base of the chimney.

David Ball Construction of Redruth became very involved in the project; they built the oak chassis and supplied the oak and ash for the wheels. The wheels were constructed by Robert Hurford, a Liveryman of the Worshipful Company of Wheelwrights at Hillfarrance in Somerset, on iron hubs cast by CompAir UK. His use of an original 19th Century sinking tyring plate to fit flaming hot steel tyres on the wooden rims was an experience that those of us who watched will never forget. The front wheels by which the vehicle is steered are 22" in diameter with a 3" tyre whilst the rear driving wheels are 33" diameter with 4" tyres. Robert Hurford also donated three original packets of Nettlefold's tyre bolts, which were purchased in 1975 as a part of a closing down sale at a wheelwrights shop in Cardiff. It turned out that these bolts were made in 1908 to repair horse bus wheels in Cardiff and were finally to be used 92 years later.

About this time a deputation from the team visited the Holman factory of CompAir UK and met Mr Geoff White, a vice-president of Lean Enterprise, a company within the Industrial Drives Division of Invensys Plc. and Mr Kingsley Rickard, the Operations Manager at Camborne.

After the visitors had explained the Society's intention to construct a replica of Richard Trevithick's 1801 road carriage Mr White enquired how CompAir might be able to help the project.

He received the reply that, had the factory retained its apprentice school its members would probably have already made considerable progress with the construction.

"Our apprentices can build it," exclaimed Mr White.

The rather puzzled visitors enquired how he saw this being achieved without apprentices.

"Oh, we've still got plenty of apprentices," said Mr White with a smile, "they're retired and wandering around Camborne, but they're still our apprentices."

The team knew then that they had the very

JOHN WOODWARD, J.W. ENGINEERING, CAMBORNE

welcome support of CompAir.

As well as being the 200th anniversary of Trevithick's first journey under steam power, 2001 is also the bicentennial year of the CompAir Holman factory in Camborne. It is interesting to speculate that Holmans is probably the oldest engineering company in the world with a continued existence in the same locality.

The subsequent building of the replica within the factory took place in an enthusiastic atmosphere of curiosity and pride. Its engineering, whilst to the highest standards and calling upon the ingenuity of all involved, was to follow the practices of two centuries ago. The CompAir engineers, used to the exacting standards employed in the construction of modern high pressure compressors, were horrified to see parts of the engine being assembled with gaps rather than tolerances. Many expressed their opinion that it would never work, an accusation John Sawle was quick to counter.

The greatest advantages of Holman hospitality were the ready access to machine tools, the skills and enthusiasm of those involved and the warm, fully equipped environment where the work could progress without delay. The Trevithick Society and the people of Camborne will always owe a great debt of gratitude to CompAir for their encouragement and unstinting generosity throughout this project.

Drawings were supplied to numerous manufacturers around Cornwall. Soon materials and parts of all sorts started to arrive at Camborne. Of particular importance were the hand forged items made by Dave Prout, a blacksmith of Pool. Using methods, which would have been recognized by Trevithick, his products added an especial authenticity to the working and appearance of the locomotive.

John was a frequent visitor at Teagle Machinery of Blackwater and Mr Fred Teagle was to become one of the most enthusiastic and generous of the suppliers.

The crosshead beam was built in the slender form of that shown on the model held by Trevithick as a statue outside the library at Camborne. To achieve the required strength the beam was cast by CompAir Reavell of Ipswich after a computerised stress analysis had been conducted by Cronite Castings of Crewkerne, Somerset. These Twenty-first Century methods would have fascinated Trevithick.

Very often a finished component was the result of co-operation between several suppliers. For example, the connecting rods were made at Pendennis Shipyard, Falmouth, from steel supplied by Macready of Newport. The bronze bearings at each end came from A. & P. Falmouth Docks, and were held in place with bolts from J. W. Engineering at Camborne. The final fitting was carried out at CompAir, Camborne.

The regular planning meetings ceased and work now commenced in earnest on the building of the replica. Hundreds of specially made square headed nuts and bolts with Whitworth threads added a visual touch of authenticity and Imperial dimensions were used throughout the machine. The engineering construction team at this stage was led by John Sawle and was based around the volunteers Arthur Young and David Bray.

During this time the team was invited to give lectures to a variety of bodies from the Truro Inner Circle to the Balwest Heritage Society. Distance was no object and Phil Hosken took the model to the United States at the invitation of Penkernewek, the Pennsylvanian Cornish Society. This trip included several days at the 1999 Gathering of Cornish Cousins at Pen Argyl and a lecture to the Blue Mountain Gas & Steam Engine Society of Bangor, PA. Lectures are booked well into 2002 and a particularly successful one was given to the Cornish Institute of Engineers in January 2001. Local pride and interest in the project continues to grow.

By the end of April 2000 the construction had reached the stage of fitting the boiler to the chassis, attaching the axles and wheels and mounting the whole replica on a low trailer loaned by CompAir to make its first appearance at Camborne Trevithick Day. The fire was lit in the furnace for the first time by Jon Eastman who went on to spend his 16th birthday tending the loco. The replica of the Camborne Car grabbed the attention of hundreds of people in the sunshine and it was clear that it had made a great impact.

JON EASTMAN ON HIS SIXTEENTH BIRTHDAY

Work continued throughout the year with a break to attend the St. Agnes Rally of the West of England Steam Engine Society. Again the loco

CHRIS TREGENZA MAKING THE CHASSIS AT DAVID BALL CONSTRUCTION, REDRUTH

was in steam, though not pressurised, and attracted a great deal of attention, including a couple who came all the way from South Africa specifically to see it. The rally coincided with the publication of *Richard Trevithick - Giant of Steam* by biographer Anthony Burton. Several boxes of the book arrived on the first day of the show and all were sold.

The team had previously received an invitation to lead the prestigious Rail 2000 Millennium event being held over August Bank Holiday on the former Stockton to Darlington railway line. The replica was to take its rightful position ahead of the world's most important locomotives. Representing the first self-propelled carriage to carry passengers it was due to ride on a low loader wagon before some 300,000 people.

After lengthy negotiations with the GOSW, permission was obtained to take the loco out of Cornwall for this particular event. The European Commission had apparently laid down rules that money, or the benefits thereof, which were intended for use in one area should not be used elsewhere. Therefore the proceeds of the Objective 5b fund for Cornwall should not be seen in County Durham. It is believed that these rules were made after Greek shipping magnates used EC cash to build ferries for use between the Greek islands which were eventually found in the Far East. Not a situation which was likely to occur in the life of the Trevithick replica!

At this time, discussions between the team and Virgin Rail had resulted in an agreement that the new Voyager Class *Cornishman Express* train to be brought into service in 2004 would be called *Richard Trevithick.* This was seen as a fitting name for an express that had run since the days of Brunel's wide gauge railway and would come into service on the 200th anniversary of Trevithick's demonstration of the first railway locomotive at Penydarren in 1804. It was also planned that a Virgin Voyager locomotive would be mocked up with the name *Richard Trevithick* and be a fitting conclusion to the Rail 2000 parade.

After all the excitement and anticipation, the Rail 2000 event ran into financial problems and was cancelled. So passed an excellent opportunity for the replica to publicise Cornwall's vision, engineering prowess and its contribution to the world's Industrial Revolution.

Whilst respect grew for Richard Trevithick for his achievements of 200 years ago many problems arose which he would not have encountered in his day. The Health & Safety Executive was not to appear for another 175 years so he had no idea of the boiler construction regulations that anyone repeating his work would face in the future. It speaks well for Trevithick's work that, developing high pressure steam engines at the very edge of the contemporary technology, he was able to design and build vessels that would withstand all the pressure he cared to exert within them. The replica uses a calibrated and weighted safety valve lever in accordance with the original design. The usual spherical weight has been incorporated on the safety valve. 200-years-ago this would have been a six-inch cannon ball.

DAVE MITCHELL AND CHARLIE DANIELS
RIVETING THE BOILER BACKPLATE AT R.K. PRIDHAM ENGINEERING

Concessions had to be made to modern practice such as the subtle fitting of a water level gauge. Trevithick had used two brass taps at different levels to ascertain the water level. These have also been faithfully replicated but are largely cosmetic. The cylinder is fitted within the boiler as Trevithick intended and has a bore of 7" and a stroke of 16". The piston is fitted with an expanding piston ring. The working crank speed will be 35 strokes per minute. This will produce a speed of 3½ m.p.h. which is intended to be suitable for the speed of the procession on Trevithick Day and other events.

Steam to the cylinder is controlled by the regulator valve which is a tapered rotary plug and a four way valve operated by a plug rod on the crosshead.

At one stage the introduction of ammonia chloride into the boiler caused consternation within the CompAir factory when a 'Hazardous to Health' label was seen on the tin. John Sawle put people's minds at rest by eating a little.

Brakes have been fitted although little thought would have been given to them in Trevithick's day. The boiler incorporates a fusible plug, a feature which at least one historian claims was fitted to the original. Trevithick found that the steering kicked violently on uneven surfaces and this characteristic has been faithfully reproduced in the replica. Suitable modifications incorporating a ball thrust bearing and the use of removable wooden pegs have greatly increased control of the steering.

During the design stage it was realised that there would be advantages in driving the loco up-hill in reverse. This would create a greater steam space below the cylinder and an increased depth of water over the furnace. The strange appearance of the loco with its chimney to the rear probably accounts for the line in the song;

'Going up Camborne Hill, coming down'

The boiler is expected to use 35 gallons of water per hour. A steam injector has been temporarily fitted for the first trials. This will be

JOHN SAWLE HAVING A RIVETING TIME

replaced by a driver controlled direct acting feed pump in gun metal driven from the crosshead as on the original. The pump with a 16" stroke and a 1" bore is capable of delivering 80 gallons per hour.

At every stage in the construction of the replica the engineers have had the advantage of the use of computer aided machine tools. This has given all concerned a greater respect for the work of Trevithick and the craftsmen of his day. They must have been working at the very edge of technology when they made the components for the very first time. Many have marvelled at Trevithick's riveted construction of the elbow bend furnace tube within the boiler. Not only did it have to contain the fire and be waterproof but it also had to withstand high pressures.

The 20" diameter furnace has a grate area of 4½ sq feet. The return flue and chimney have a diameter of 12". The chimney has an internal metal liner and the top section is removable.

Truro Tractors supplied 750 rivets. The 36 gallon feed water tank was made of 25 pieces of steel held together by 270 hot rivets. The tank and many of the external water pipes have been galvanised by Cornwall Galvanising to ensure that the loco will last for at least 100 years. The remainder of the rivets were used in the construction of the chimney.

Lubrication has received considerable attention. All bearings, except the 30" flywheels which will turn at three times the crankshaft speed and have disguised grease nipples, are supplied with oil through siphon cups. The cylinder uses a displacement lubricator with traditional lard oil.

The name, age, town and trade of every individual who has worked on the replica has been stamped into the metal parts under the loco. The oldest person to make parts was Fred Harris, 85, of Teagle Machinery and Holman's contribution has been enshrined for all time by a contribution of entrepreneurial and engineering skills from the well-known Jan Luke. His presence at projects not directly related to company production has been appreciated over the years by all that have worked at Holman's.

As the construction neared completion the replica underwent a period of testing. It was initially operated under compressed air within the factory prior to handling and speed trials, which were under steam, along the very suitable service roads to be found at the rear of the Holman factory. Again, it is impossible to over emphasis the contribution which has been made by CompAir at their Holman factory.

All those who have contributed in anyway to the completion of this fine vehicle will feel proud when it leads the Trevithick Day parade for the first time on the 28th April 2001. Engineers who have seen it being built have remarked upon the excellent standard of workmanship which has been employed at all stages of its construction.

As news of the replica reached around the globe the team have received numerous enquiries from model engineers who wished to build a scale replica themselves. The provision of these facilities has had to take second place behind the construction of the full-size Camborne engine but arrangements are being made with a precision engineering company, well known for the quality of their work, to market drawings and parts for model engineers.

In conclusion, all those that have been involved in this project remain very much in awe of Richard Trevithick and his co-workers who with the most primitive of manufacturing facilities crafted and successfully operated the engine that changed the pace and direction of the Industrial Revolution and ultimately powered and propelled the emerging modern world.

The people of Cornwall, and Camborne in particular, can feel especially proud of the achievements of Richard Trevithick and the members of the Society named after him. Trevithick was a man of vision who was always looking forward. In this respect he saw a great deal that many of his unconvinced contemporaries were unable to comprehend. A frustrated Trevithick saw

DAVID BRAY AND JOHN SAWLE ASSEMBLING THE BOILER

his boiler and steam engine powering the world's industry and transportation. Today we must wonder if even Trevithick was able to visualise the tremendous contribution his practical and exploratory work in high pressure steam was going to provide.

Two hundred years on we can look back at the railways which have spanned the globe, the mighty ocean going liners and the multiple uses for steam in industry. Trevithick's containerisation of cargo and the fitting of steam engines to ships in his time must have given him a glimpse into a future. A future where the world changed more quickly than at any other time.

The development of mechanised industry and self-propelled transportation all started in Camborne two centuries ago with a public demonstration of high pressure steam when Richard Trevithick drove,

'Up Camborne Hill'

TEAGLE MACHINERY LTD, BLACKWATER

KEN VINCENT, A. & P. FALMOUTH DOCKS

ARTHUR YOUNG WORKING ON THE CYLINDER BLOCK

Some of the inventions and activities of Richard Trevithick.	
1771	Born in the parish of Illogan.
1797	Marriage to Jane Harvey. Death of father.
1798	The return flue boiler and the high pressure engine.
1801	The Camborne road locomotive.
1802	Patent for High Pressure Engine granted.
1803	The London Road Carriage - the first coach (replica made). Boiler explosion at Greenwich. The Coalbrookdale Locomotive - the first portable engine to run on rails (actually a plateway). Replica made.
1804	The Penydarren Locomotive - the first train and the first engine to pull more than its own weight. Replica made.
1805	The Wylam locomotive, Newcastle-on-Tyne. Drove a barge by steam engine and paddle wheels.
1806	Applied his engine to a dredger.
1807	Appointed engineer to the Thames Archway Company—the Thames Tunnel scheme.
1808	Catch-me-who-can railway locomotive - the first fare-paying passenger train. Containerisation of ships cargoes.
1809	Demonstrated that a ship could be made of iron. Raised a sunken ship off Margate.
1810	Typhus and return to Cornwall.
1811	Trevithick bankrupt. First Cornish engine and Cornish boiler.
1812	Trewithen threshing machine - using his engine for agriculture. Rock-boring machine for Plymouth breakwater.
1813	Meets Francisco Uvillé.
1815	Plunger pole engine. A ship's propeller.
1816	Sails for Peru from Penzance.
1818	First steamboat fitted with Trevithick puffer.
1821	Salvages sunken cargo near Callao, South America.
1827	Creates recoil gun carriage for Bolivar.
1828	Visits Holland. Engine and pump made in Hayle for Zyder Zee.
1829	Patent for water-jet propulsion and room heater.
1832	Tubular superheated boiler. Proposed design for 1000' cast iron tower to commemorate the Reform Bill.
1833	Died in Dartford, Kent.

Richard Trevithick also invented:
The multi-tubular boiler, the fusible plug, the exhaust steam blast, the kibbal, the water pressure pumping engine,
The recoil engine, the rotary plough, the plunger pole pump and refrigeration.

Thanks are due to the following companies and individuals who provided all or part of their services and/or supplies free of charge.

A. & P. (Falmouth) Ltd	Falmouth Docks	Forging, non-ferrous castings, machining axles and other components.
S. J. Andrew & Son	Redruth	Supplies of steel.
Andue Crafts	Newlyn East	Woodturning
Compair UK – Holmans Compair Reavell	Camborne Ipswich	Pattern making, iron casting, supplies of steel and bronze, machining of components, workshop space and facilities for assembly and fabrication.
David Ball Construction Ltd (part of Rowe Holdings)	Redruth	Manufacture of oak chassis, supply of oak & ash for wheels.
Beldam-Lascar Ltd	Falmouth	Boiler joints and gaskets.
Brownnall Nabic (NVSP Ltd)	Manchester	Fusible plugs.
BSL Bearings Ltd	Redruth	Ball thrust bearing for steering.
Cerdic Foundry Ltd	Chard	Bronze castings.
Cronite Castings Ltd	Crewkerne	Stress analysis.
John Crowther, Design Bureau	Truro	Printing drawings.
Deepdale Engineering Ltd	Dudley	Boiler construction and chimney elbow.
Duchy Plastics	Camborne	Rust proofing components.
FWB Southwest Ltd	Threemilestone	Stainless steel nuts and bolts.
Graham Builders Merchants	Pool	Flue liner.
Robert Hurford	Taunton	Fabricating wooden wheels.
International Coatings Ltd	Gateshead (Falmouth Depot)	Specialist paint.
J.W. Engineering	Camborne	Machining nuts, bolts and studs.
Robert C. Keat	Carclew	Stainless steel components.
Kernow Oils	Penzance	Lubricants.
Maker Coating Systems Ltd	Newton Abbot	Paint and rust preventative treatments
MPS Ltd	St. Agnes	Shot blasting.
Macready's Ltd	Newport	Steel supplies.
Pendennis Shipyard Ltd	Falmouth Docks	Manufacturing connecting rods.
Pressvess Process Engineering	Kingswinford	Angle ring.
Pridham Engineering Ltd	Tavistock	Riveting boiler components.
David Prout	Pool	Blacksmithing, forging.
Qprint	Camborne	Printing drawings.
RNAS Culdrose	Helston	Forging and fabrication.
Seetru Ltd	Bristol	Two cylinder relief valves.
SPF Inspection & Consultancy	Arundel, Sussex	Boiler inspection.
Truro Tractors Ltd	Chacewater	Supply of rivets.
Teagle Machinery Ltd	Blackwater	Supply of steel, profile cutting and machining.
Terrill Bros. (Founders) Ltd	Hayle	Casting in grey & S.G. iron.
Nigel Watts	Penponds	Patternmaking.
Westfield Transport	Chacewater	Transport of boiler & wheels.

Thanks are due to the following who provided financial support to the project.
EC Konver II
Kerrier District Council
The Tanner Trust
Camborne Town Council
The Trevithick Society
Prosper
West of England Steam Engine Society
Frank Trevithick-Okuno

PAUL DOBLE, RONNIE HIGMAN AND DAVID BRAY
BOLTING DOWN THE BOILER IN READINESS FOR TREVITHICK DAY 2000
A MAJOR STEP FORWARD IN THE CONSTRUCTION OF THE REPLICA

FURTHER READING

Burton, A. 2000. *Richard Trevithick, Giant of Steam*. Aurum.
Dickinson, H.W. & Titley, A. 1934. *Richard Trevithick, the Engineer and the Man*. Cambridge Univ. Press.
Eyles, J.M. 1971. *William Smith, Richard Trevithick and Samuel Homfray. Their correspondence on steam engines.* Institution of Mechanical Engineers.
Farey, J. *Treatise on the Steam Engine*, Vol. 2. David & Charles, 1971.
Hodge, James. 1973. *Richard Trevithick*. Shire Publications.
Inglis, C.E. 1933. *Trevithick Centenary Commemoration Memorial Lecture*. Institution of Civil Engineers.
Odgers, J. F. 1950. *Richard Trevithick "The Cornish Giant" 1771-1833. Memorials and Commemorations in Camborne and District.* Camborne Printing and Stationery Co.
Rolt, L.T.C. 1960. *The Cornish Giant*. Lutterworth Press.
Trevithick, F. 1872. *Life of Richard Trevithick*. E. & F.N. Spon.
Twining, E.W. 1938. *The First Railway Locomotive Engine*. The TTR Gazette. Nos 3 & 4.
Woodcock, Lloyd H. 1971. *Richard Trevithick's First Steam-locomotive Trial: Christmas Eve, 1801*. Camborne Festival Magazine, pp. 9-15.
Also Trevithick Society Newsletters 92 to 112 which included details of the 2001 Project.

CHRIS BRANWELL MACHINING THE BRONZE BUSH ON A BORING MILL, HOLMAN FACTORY, COMPAIR

"I have been branded with folly and madness for attempting what the world calls impossibilities, and even from the great engineer, the late Mr. James Watt, who said to an eminent scientific character still living, that I deserved hanging for bringing into use the high-pressure engine. This so far has been my reward from the public; but should this be all, I shall be satisfied by the great secret pleasure and laudable pride that I feel in my own breast from having been the instrument of bringing forward and maturing new principles and new arrangements of boundless value to my country. However much I may be straitened in pecunary circumstances, the great honour of being a useful subject can never be taken from me, which to me far exceeds riches".

Richard Trevithick.
1771-1833